Ten Elderly Victims from
Intermediate Care Facility Fire
Colorado Springs, Colorado

Investigated by: Jack Yates

This is Report 050 of the Major Fires Investigation Project conducted by TriData Corporation under contract EMW-90-C-3338 to the United States Fire Administration, Federal Emergency Management Agency.

 FEMA

Department of Homeland Security
United States Fire Administration
National Fire Data Center

U.S. Fire Administration Fire Investigations Program

The U.S. Fire Administration develops reports on selected major fires throughout the country. The fires usually involve multiple deaths or a large loss of property. But the primary criterion for deciding to do a report is whether it will result in significant "lessons learned." In some cases these lessons bring to light new knowledge about fire--the effect of building construction or contents, human behavior in fire, etc. In other cases, the lessons are not new but are serious enough to highlight once again, with yet another fire tragedy report. In some cases, special reports are developed to discuss events, drills, or new technologies which are of interest to the fire service.

The reports are sent to fire magazines and are distributed at National and Regional fire meetings. The International Association of Fire Chiefs assists the USFA in disseminating the findings throughout the fire service. On a continuing basis the reports are available on request from the USFA; announcements of their availability are published widely in fire journals and newsletters.

This body of work provides detailed information on the nature of the fire problem for policymakers who must decide on allocations of resources between fire and other pressing problems, and within the fire service to improve codes and code enforcement, training, public fire education, building technology, and other related areas.

The Fire Administration, which has no regulatory authority, sends an experienced fire investigator into a community after a major incident only after having conferred with the local fire authorities to insure that the assistance and presence of the USFA would be supportive and would in no way interfere with any review of the incident they are themselves conducting. The intent is not to arrive during the event or even immediately after, but rather after the dust settles, so that a complete and objective review of all the important aspects of the incident can be made. Local authorities review the USFA's report while it is in draft. The USFA investigator or team is available to local authorities should they wish to request technical assistance for their own investigation.

This report and its recommendations were developed by USFA staff and by TriData Corporation, Arlington, Virginia, its staff and consultants, who are under contract to assist the USFA in carrying out the Fire Reports Program.

The USFA greatly appreciates the cooperation received from the Colorado Springs Fire Department, with particular thanks to Fire Chief Louis A. Roman, Deputy Chief Terry Q. Gladdin, Captain Michael R. Gower, and Lieutenant Dave L. Stephenson. Mr. Paul Daraghy, Director of the State of Colorado Health Department also provided used information.

For additional copies of this report write to the U.S. Fire Administration, 16825 South Seton Avenue, Emmitsburg, Maryland 21727. The report is available on the USFA Web site at http://www.usfa.dhs.gov/

U.S. Fire Administration

Mission Statement

As an entity of the Department of Homeland Security, the mission of the USFA is to reduce life and economic losses due to fire and related emergencies, through leadership, advocacy, coordination, and support. We serve the Nation independently, in coordination with other Federal agencies, and in partnership with fire protection and emergency service communities. With a commitment to excellence, we provide public education, training, technology, and data initiatives.

 FEMA

TABLE OF CONTENTS

Ten Elderly Victims from Intermediate Care Facility Fire
Colorado Springs, Colorado
March 1991

Local Contacts: Louis A. Roman, Fire Chief
Terry Q. Gladdin, Deputy Chief
Colorado Springs Fire Department
31 South Weber Street
Colorado Springs, CO 80903

Michael R. Gower, Captain
Fire/Arson Investigations
Colorado Springs Fire Department
105 East Vermijo, Suite #300
Colorado Springs, CO 80903

David L. Stephenson, Lieutenant
Fire Prevention Bureau
Colorado Springs Fire Department
101 West Costilla Street
Colorado Springs, CO 80903

Paul Daraghy, Director
State of Colorado Department of Health
4210 East 11th Avenue
Denver, CO 80220

OVERVIEW

An accidental fire determined to have originated above ceiling level occurred in the early morning hours of March 4, 1991, to the Crystal Springs Estates personal care boarding home in Colorado Springs, Colorado. All indicators point to this fire as having been burning in the attic undetected for some period of time, perhaps as much as an hour or more before breaking through the ceiling drywall and having smoke detectors react.

Heat detectors were found in the attic but had not functioned. A pull station's use and a 9-1-1 call alerted the fire department, but upon arrival they found a fire already ventilated through roof turbines. Nine elderly occupants perished, all as a result of carbon monoxide poisoning, in this fire. A tenth victim died months later of smoke aggravated lung problems.

SUMMARY OF KEY ISSUES

Issues	Comments
Cause of Fire	Undetermined; first thought to be ceiling exhaust fan, but eliminated by investigation.
Fire Origin	Ceiling space of east wing.
Delayed Detection	Heat detectors in attic had not been connected or did not function. After considerable fire development, ceiling smoke alarm activated, nurse operated alarm pull station that closed fire doors and alerted central monitoring agency. Owner and nurse/director pounded on doors to wake residents. Owner called 9-1-1.
Building Structure	Single-story, brick veneer; four connected wings. Attic of east wing had breech in draft stop. Between west and north wings, fire door on first floor was offset 36 inches from firewall in attic.
Residents	Twenty-five at time of fire, aged 72 to 97; many required wheelchairs or walkers. Only two staff were present when fire occurred.
Casualties	Nine fatalities during fire; one additional fatality months later from smoke aggravated lung problems; average age 85.
Fire Protection Features	No sprinklers. Rate of rise heat detectors in halls and each room; smoke detectors in halls and dining area; hard-wired, no battery backup. Fire doors operated.
Inspections & Compliance	State inspections incomplete due to limited staff but facility licensed to operate. Classed as "intermediate or personal care facility" rather than nursing home, which has more stringent fire protection requirements.

STRUCTURE

This facility was one constructed in three different phases. The first construction plans included the west and south wings which were begun in 1959. (See Appendix A for floor plan.) The east wing was added in 1960 and the north wing in 1962. The north and east wings do not connect and an opening large enough for vehicles to drive through was left at the northeast corner allowing access to a courtyard area in the center of the complex.

The exterior of this structure was of brick veneer. The interior finish consisted of drywall ceilings and walls. The floors were concrete slab with covering of carpet in some areas and tile in others.

All occupants' rooms exited onto hallways that would lead to exterior exits and all rooms had windows approximately 48 inches above floor level. The facility appeared to comply with the National Fire Protection Association (NFPA) Life Safety Code requirements that apply to evacuation.

However, the actual structure did not have safety features that originally were thought to be there; in particular, properly rated and designed firewalls. A wall at the east end of the dining area was thought at first to be a firewall, however, closer inspection revealed that a tongue-in-groove ceiling extended over the top leaving exposed wood to the outside. A cinder block wall was constructed adjacent to it, but the integrity of the wall was breached when holes were punched in, in order to allow the ceiling joist for the east addition.

The second firewall was thought to be at the area of the addition of the north wing where it tied into the existing west wing. A closer inspection showed that the firewall in the attic was not designed to extend to floor level but in fact, the fire door for this wing was set approximately 36 inches north of the attic stop, again rendering this to be less than a fully rated firewall.

In this fire, this attic barricade along with the drywall ceiling did impede the communication of the fire in the west wing and prevent travel of the fire into the north wing area. However, had the fire begun closer to this wall, it most probably would not have stopped fire extension as it was thought to be designed for.

THE FIRE

At 0038 hours on March 4, 1991, the Colorado Springs Fire Department received an alarm notification of a structure fire at 825 South Hancock which is the address for Crystal Springs Estate. The first notification came from an alarm monitoring company out of Denver that monitored the Crystal Springs Estate system. The alarm was activated by a pull station in the facility. Within seconds of the alarm company's call to the fire department, a second call, on 9-1-1, was received verifying actual fire at the scene. (See the Colorado Springs Fire Department's Report on the Crystal Springs Estate Fire in Appendix B for further details on the early stages of the incident and the investigation findings.)

The initial alarm recommendation of one company was upgraded to two engine companies and one truck company. Upon arrival the first responding company reported visible fire and immediately requested more units to respond to assist in evacuation and suppression.

The fire department began immediate rescue operations and assumed an offensive fire attack. Occupants and victims were removed and the fire department attacked the fire for over an hour before they were forced to pull personnel from inside the building and then assume a defensive position.

Eventual damage was extensive causing total destruction to the roof over the east, south, and most of the west wing. The north wing still had roof area intact. Heavy interior damage occurred to the east, south, and west wings as a result of fall down and secondary burning.

A total of 25 residents were in the structure at the time of the fire along with two staff members. Nine fire fatality victims were removed during the suppression of the fire. The victims ranged in aged from 77 to 97. The El Paso County Coroner's Office has determined that all the victims died from asphyxia due to carbon monoxide inhalation. All victims but one were found in their rooms; the one outside of her room was found in the dining area. One woman, Lois Mitchell, age 82, died in December 1991 of smoke aggravated lung problems.

In addition, eight residents of the home were treated for smoke inhalation and five firefighters were injured.

The people who died from this tragic fire were:

Name	Age
Ruby Bowes	97
Ethyl Haythorn	90
Louise Evans	88
Velma McKahan	88
Muriel Hartman	84
Miriam Wilson	82
Lois Mitchell	82
Marion Staples	81
Catharine Martin	78
Alice Shuey	77

CODES AND INSPECTIONS

At the time this property was constructed, there were no codes that would have prescribed installation of sprinkler systems. In Colorado, the intermediate care facilities such as this one are licensed and inspected by State Health Department Inspectors who work with NFPA Life Safety Code 101. Colorado Springs, however, subscribes to the Uniform Fire Code (UFC) as their regulatory guide.

Since this building was constructed it has been used for multiple purposes. The current owner applied for his license midyear of 1988. The three initial inspections are conducted by the fire department, zoning, and then building inspectors. The State Health Department then sends in their inspector. The first State Health Inspector was at the facility in August of 1989. The director of Colorado Department of Health indicated there were three additional inspections: November 27, 1989, May 16, 1990, and November 14, 1990. As of the last inspection, he stated the installation of additional smoke alarms is all the facility needed in order to pass their requirements and be licensed.

The Colorado Springs Fire Department also continued to inspect these premises under the UFC. This is listed as an I inspection (Institutional facility), and inspections are conducted by engine company personnel after the initial inspection on a biannual basis.

FIRE AND SMOKE DETECTION EQUIPMENT

This facility was equipped with rate of rise heat detectors in each room and in the halls. Smoke detectors were in-place in the halls and dining area. There were also heat detectors found within the attic space, however, in that the fire burned undetected within the attic before breaking through the ceiling, it is felt these detection devices were not connected or properly functioning.

It also appeared all systems were hard-wired and were not battery backed.

There was no sprinkler protection in the facility.

FIRE DEPARTMENT

Colorado Springs is approximately 181 square miles in size with a population of 289,000. There are two fire districts, 15 stations, 20 fire companies, and approximately 375 personnel. Within that number is the fire prevention division, which includes Code Enforcement and Fire Investigations. In addition, a public education specialist is in a full-time capacity with the department.

The Colorado Springs Fire Department is an extremely professional organization with a well charted sense of direction. Their mission statement as outlined in their 1991 goals and objectives states their overall attitude: "To promote life safety and property conservation for the citizens of Colorado Springs through compassionate, professional service."

CONTRIBUTING FACTORS REGARDING THE FIRE

This tragic fire, like many other similar fires in care facilities, illustrates common, preventable problems that continue to occur.

The first and foremost question deals with the inspection force and who is ultimately responsible for the well-being of the occupants in this type of facility. Colorado labeled this facility as an "intermediate or personal care facility" instead of calling it a "nursing home." For this category there is supposed to be a lesser degree of care needed, i.e., less nursing staff. However, as shown in this scenario, this group of residents who averaged 85 years of age, many with limited mobility, were put at a serious disadvantage by this categorization when an emergency situation arose. Other States, like Colorado, need to review their criteria and practices in this important area.

Adequacy of inspections is another problem area spotlighted by this incident. Colorado currently has two inspectors for the entire State and 281 operational personal care homes. They range in size from three beds to several hundred, but most are twenty beds or smaller. The State Health Department Director is currently trying to arrange contracts with local county inspectors to maintain inspections on these facilities, but in the interim two inspectors are charged with the major portion of inspections in the State. The director indicates these are not trained building inspectors and have only a cursory amount of training in this area.

The State Health Department Director makes the point that the bulk of these facilities are not in-line for any Federal aid and if forced to go with more stringent safety features, they would be forced to close. By comparison, numerous States, such as Virginia and Alabama, have made sprinklering of this type of facility mandatory, and it does not appear that financial constraints have hampered them.

The Colorado Springs Fire Department is to be commended for their openness during USFA's investigation of this fire and their willingness to learn from this loss. They indicated they will implement into their plan a program of more stringent code enforcement as a result of lessons learned from this fire.

The Colorado Springs Fire Department did a building inspection and outlined areas needing to be corrected, to which the owner complied. Their inspection was followed by the State Health Inspector and any additional fire safety and physical plant inequities found were to be corrected before a license to operate would be issued. The licensing process with the State began in August 1989 and, as of the date of the loss, the final inspection had not been made. The last inspection on November 14, 1990, required that an additional 12 smoke detectors be installed in the dining area and halls. That was completed approximately one week prior to the fire, so, technically, the facility met with all licensing requirements.

Yet, according to the owner, no inspector reviewed evacuation procedures or checked emergency lighting or heat detection devices in the attic space. In this fire, the fact that fire burned in the attic undetected for an extended period means greater possibility for electric service to be severed. The value of battery back-up detection systems is clear from this incident. Also, as Colorado Springs Fire Department Deputy Fire Chief Terry Gladdin stated, sprinklers would have gone a long way toward controlling the fire.

LESSONS LEARNED

1. **Inspection for these types of care facilities must be conducted by full qualified building inspectors as well as by health inspectors, or joint inspections should be conducted.**

   There is little accomplished if buildings or institutions pass one area but fail in another. The inspection process must be a comprehensive operation and completed in full.

2. **Properly working heat and smoke detection equipment is a must.**

   All detection equipment in-place must be kept in working condition. The system must be checked on a scheduled basis by trained personnel. Battery assistance to hard-wired heat/smoke detectors offers the greatest safety. Although not a factor in this loss, the scenario of this incident illustrates the threat of loss of power.

3. **Adequate staff must be assigned for the number of occupants.**

    At the time of the fire only two staff were on the premises. Many of the "intermediate care" occupants present required wheelchairs, walkers, oxygen, and other aids. In a dire emergency such as this, that was entirely too few staff people to assist the occupants.

4. **Care facilities should be sprinklered in patient rooms and halls.**

    Tragedies such as this fire demonstrates the increasing need for more individual facilities and communities to install sprinklers for the protection of life and property despite economic barriers.

5. **Evacuation planning and regular drills are particularly important in facilities for the elderly.**

    Review of this fire once again reminds us of the difficulties involved in protecting the elderly from fire. In all facilities for the elderly, but particularly residences, evacuation planning and regular drills are an essential part of responsible facility management.

APPENDIX A

Floor Plan Showing the Point of Origin, Locations of Bodies, and the Extent of the Fire

CRYSTAL SPRINGS ESTATES
Intermediate Care Facility
Colorado Springs, Colorado

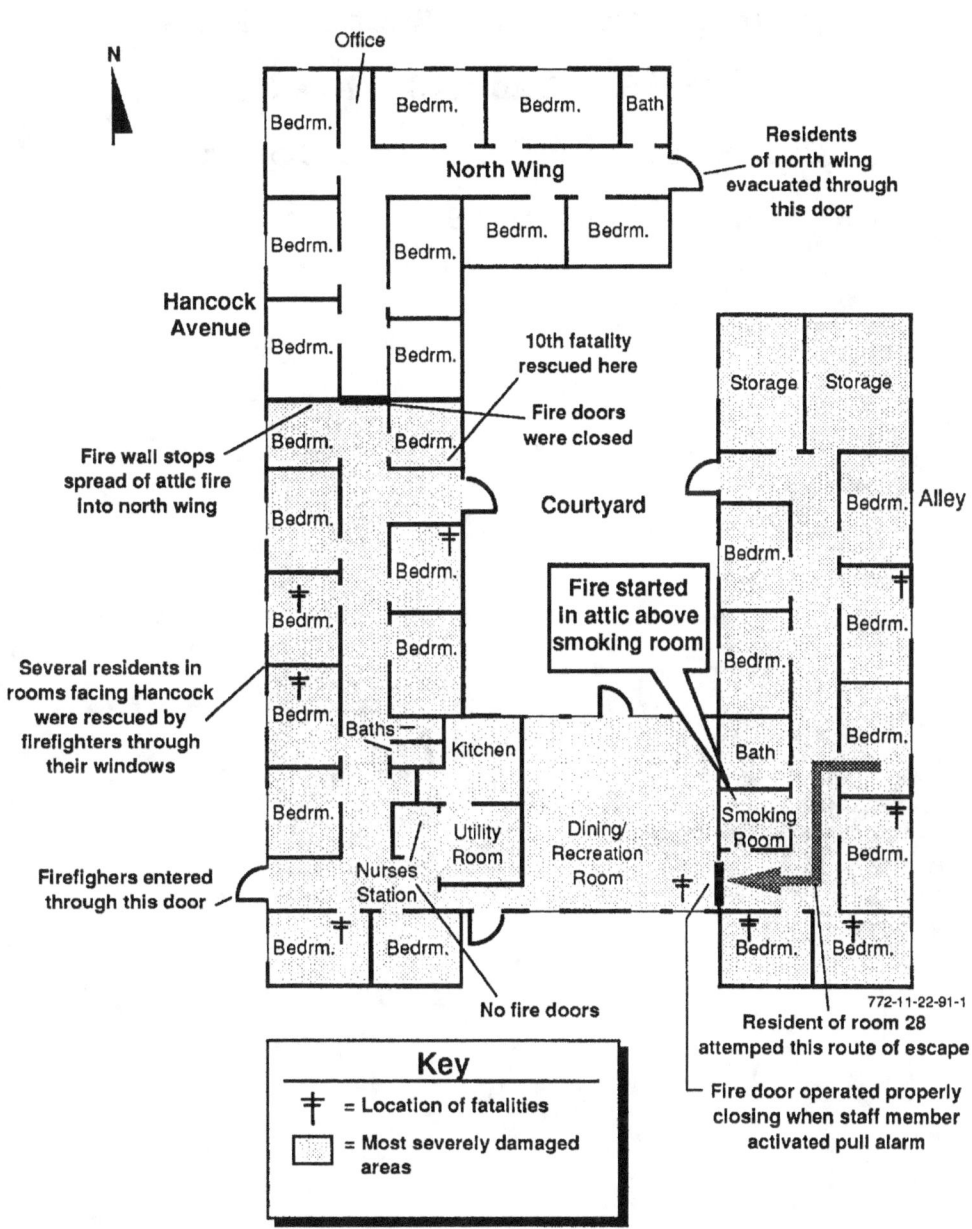

N

Office

Bedrm.

Bedrm.

Bedrm.

Bath

Residents of north wing evacuated through this door

North Wing

Bedrm.

Bedrm.

Bedrm.

Bedrm.

Hancock Avenue

Bedrm.

Bedrm.

10th fatality rescued here

Storage

Storage

Fire wall stops spread of attic fire into north wing

Bedrm.

Bedrm.

Fire doors were closed

Bedrm.

Alley

Bedrm.

Courtyard

Bedrm.

Bedrm.

Fire started in attic above smoking room

Bedrm.

Several residents in rooms facing Hancock were rescued by firefighters through their windows

Bedrm.

Bedrm.

Bedrm.

Baths

Kitchen

Bath

Bedrm.

Firefighters entered through this door

Bedrm.

Utility Room

Dining/ Recreation Room

Smoking Room

Bedrm.

Nurses Station

Bedrm.

Bedrm.

Bedrm.

Bedrm.

Bedrm.

No fire doors

772-11-22-91-1

Resident of room 28 attemped this route of escape

Fire door operated properly closing when staff member activated pull alarm

Key

✝ = Location of fatalities

☐ = Most severely damaged areas

7

APPENDIX B

Colorado Springs Fire Department Report on the Crystal Springs Estate Fire

```
Crystal   Springs   Estate   Fire
       825  S.  Hancock  Ave.
   Colorado   Springs,   Colorado
              3/04/91
```

Appendix B (continued)

Crystal Springs Estate Fire
825 S. Hancock Ave., Colorado Springs, Colorado
3/04/91

This fire occurred in a "Personal Care Boarding Home" with 24 residents and 2 staff members in the building, 34 minutes after midnight, on a Monday morning.

STRUCTURE:

The building is a single-story, brick veneer structure. It is configured as 4 interconnected wings which enclose a center courtyard. The north and east wings do not connect at the northeast corner of the property, allowing access to the center courtyard. The sketch below illustrates the general layout of this 11,590 square foot building.

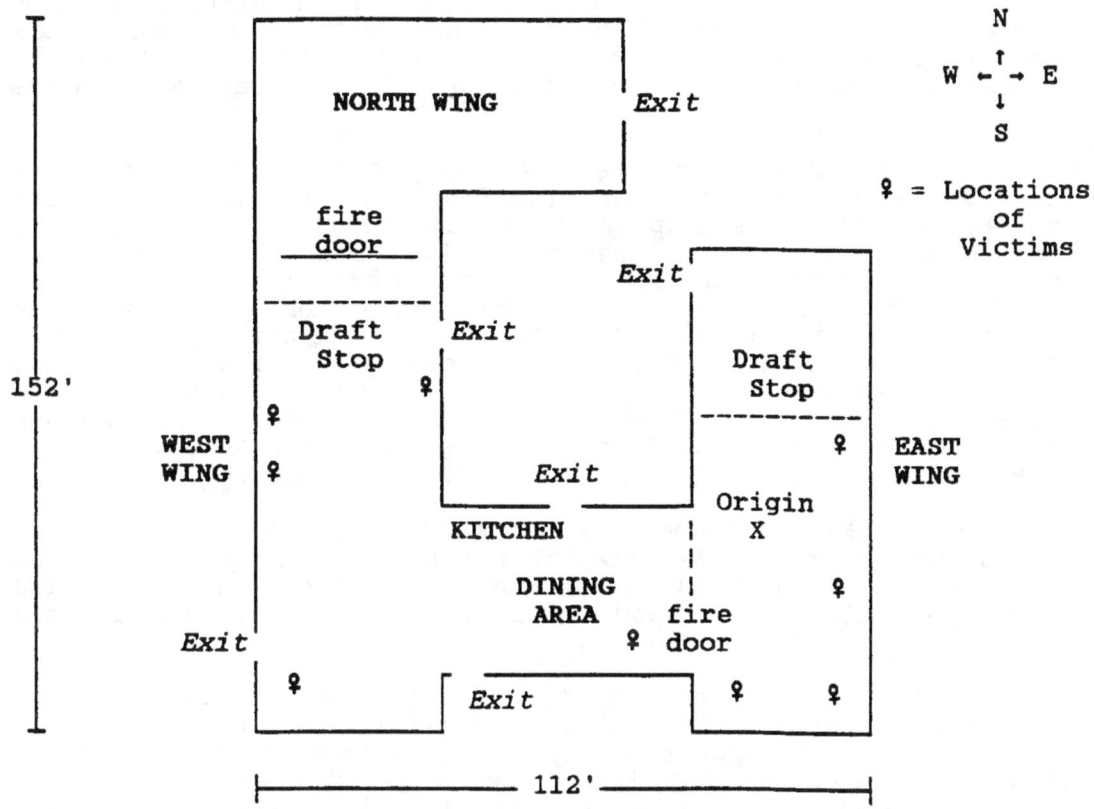

The east wing contains 8 resident rooms and a smoking room. There is a draft stop separating the north one-third of the attic from the larger south portion. This draft stop, however, is breached by an opening measuring approximately 4' by 4' according to an electrician performing work in the building the week before the fire.

Rev. 3/21/91

Appendix B (continued)

Crystal Springs Estate
3/04/91   Page 2

The east wing connects to the dining and kitchen area (south wing) at its southern end.   A single, automatic closing fire door separates these two wings. Firefighters conducting a search of the building found it in the closed position during the fire.

The west end of the dining room connects to the west wing, which contains 10 resident rooms and the nurses station. No fire doors or walls separate the south wing from the west wing.

The north wing contains 9 resident rooms and the business office. The north and west wings were originally thought to be separated by a brick fire wall and automatic closing door. A closer inspection revealed that the rated doorway in the hall is positioned approximately 4' north of the brick fire wall. This reduces the protection provided to nothing more than a brick draft stop in the attic.

Construction plans indicate the west and south wings were constructed in 1959, the east in 1960, the north in 1962.   The entire facility is equipped with heat detectors in the resident rooms, hallways, and attic spaces.   Pull stations, a few smoke detectors, and portable fire extinguishers are also located in the hallways of the facility. A post-fire inspection of the remaining systems and wiring indicated the pull stations, older smoke detectors, and heat detectors were connected in series on a 2-wire loop. The activation of a pull station or a detector was designed to send an alarm signal to the central monitoring agency. Neither the management of the facility nor the alarm monitor could verify the condition of the alarm system prior to the fire.

An electrician who inspected this system stated that if a fire burns through one of the wires in the loop, every portion of the loop beyond that point would become deactivated. Also, a 2-wire system is not able to send a "trouble" indication should this scenario take place.

To satisfy the requirements of the State Health Department, an electrician installed 12 additional smoke detectors in the dining room and hallways approximately 1 week before the fire.   These ionization detectors were interconnected so the activation of any detector would cause all 12 to sound an audible alarm. This 110 volt system received power through a separate circuit and was designed to sound a "local alarm" only. After verification of an emergency, a staff member must activate a pull station to send an alarm to the alarm monitor.

Rev. 3/21/91

Appendix B (continued)

Crystal Springs Estate
3/04/91   Page 3

OCCUPANCY AT THE TIME OF THE FIRE:

As a "Personal Care Boarding Home," Crystal Springs Estate provided meals, lodging, entertainment, and limited medical assistance to 25 residents ranging from age 72 to 97. Many required walkers and wheelchairs to move about. One person who resided at Crystal Springs was already at a hospital when the fire occurred. Everyone except the occupant of room 28 was asleep in bed when a staff person made her midnight rounds. This occupant was later found in the dining area, fatally injured by the fire.

Two Staff members were in the building when the fire was discovered. One, an owner of the facility, was asleep in the north wing. The other, a registered nurse and the Director of the home, was setting out silverware in the dining area in preparation for breakfast.

ACTIVITIES AT THE TIME THE FIRE WAS DETECTED:

The Director of the facility reported that she was working in the dining room when the "local" fire alarm system activated. This employee also heard at least 1 person in the east wing screaming "fire." The audible alarm most likely was activated by the smoke detection system which was recently installed.

The Director stated she went into the east wing, looked north, and observed a considerable amount of fire coming out of the smoking room doorway. She then ran to the southwest corner of the dining room, near the exit, and operated an alarm pull station. The fire doors in the building immediately closed and an alarm was sent to the central monitoring agency in Denver, Colorado at 12:34 a.m. She continued to run through the building to the north wing, where she awoke the other staff member. Together they pounded on the doors of the north and west wings to wake the residents.

Upon arriving where the west wing meets the dining room, the staff members found the dining room to be well involved in fire. The owner then called the 911 emergency center from the nurses station at that location. This was recorded as 12:37 a.m., which would be approximately 3 minutes after the pull station was operated by the Director.

At 12:42 a.m. the first-due engine company arrived and found fire to be ventilating from the north windows of the dining room and the roof above the south end of the east wing. Fire suppression and rescue activities were begun immediately.

Rev. 3/21/91

Appendix B (continued)

Crystal Springs Estate
3/04/91   Page 4
─────────────────────

INVESTIGATION FINDINGS:

The fire destroyed the entire east wing, the kitchen and dining
facilities, and the south two-thirds of the west wing. The fire
did not travel north of the brick wall which separates the attics
of the west and north wings.

The smoking room is located on the west side of the east wing,
directly abreast of the dining room. Examination of the area above
the smoking room indicated the fire originated above the ceiling
and spread throughout the attic of the east wing.

A "consumer grade" 110 volt ceiling exhaust fan, similar to that
found in residential bathrooms, was recovered from the debris in
the smoking room. An employee informed investigators that the
ceiling fan was controlled by the overhead lighting and was very
noisy. She also said the fan was present when the building was
purchased by the current owners.

Closer inspection revealed that the bearing assembly to the fan
showed evidence of extreme wear, with the bearing/bushing being
worn into an oblong shape.   An owner later stated that
approximately 1 month before the fire, the exhaust fan became very
noisy so someone removed the fan cover and sprayed WD-40 into the
motor housing. This reduced the amount of noise the fan made.

The double 2"x 4" top plate on the north wall of the smoking room
showed evidence of extreme heat and near total destruction above
the location where the ventilating fan was found. Distinct burn
patterns were also found on the wall beneath this area of the
room's north wall. A wall mounted cabinet was also burned at its
top, east edge immediately adjacent to the exhaust fan area.

An electrician inspected the circuit to the overhead lighting and
exhaust fan and found them to be powered by the only 15 amp circuit
breaker in the electrical panel. The breaker was tripped sometime
during the fire and showed evidence of extreme internal heat. The
electrician stated the breaker was no longer functional and
indicated that it may have been operating very near its capacity
for a long time. Adjacent breakers showed no signs of heat and
appeared to remain functional.

A second area of severe burning was located approximately 8' west
of the exhaust fan, at the enclosure to the furnace flue extending
from the basement. This flue is immediately next to the extreme
east end of the dining room. The top plate surrounding this flue
revealed extreme burning on the north side where a support strap

Rev. 3/21/91

Appendix B (continued)

Crystal Springs Estate
3/04/91   Page 5

---

connects to a ceiling joist. This was originally thought to be the
origin of the fire.   It was during excavation of the smoking room
that the ceiling fan and its nearby structural members were
discovered.

A representative of the National Fire Protection Association, upon
viewing the two areas of burn, theorized that the fire burned
deeply near the flue because   the enclosure was open and
unrestricted from the basement to the attic.   This permitted a
sustained flow of fresh air into the attic at this point, which
allowed very hot, and efficient, burning to take place.

The fire originated near the ceiling exhaust fan in the smoking
room of the east wing and extended west into the dining room by way
of a void between the cinderblock and brick veneer of the dining
room's east wall.   Openings were made in the brick veneer to
accommodate the mounting of roof members for the east wing.

The dining room burned at a very high temperature due to its heavy
timber roof/ceiling construction and possible cooking grease/oil
deposits which had adhered to it over time. Fire further extended
into the west wing, burning north until being stopped by the brick
wall subdividing the attic space.

It is unknown why the heat detectors found in the attic area of the
east wing did not warn of the fire in that area. Inspection of the
alarm panel revealed 2 fuses which were blown at some unknown time.
The condition of the system before   the fire has not been
established.

Investigators feel that the fire burned in the attic above the
residents, undetected, for a considerable time.   Only when the
ceiling in the smoking room collapsed, did fire products reach the
recently installed smoke detector system and activate a local
alarm.

Rev. 3/21/91

Appendix B (continued)

Crystal Springs Estate
3/04/91   Page 6
_____

FIRE COSTS:

Nine residents were found overcome within the structure and were
pronounced dead by the El Paso County Coroner's Office. Autopsies
revealed that all were killed by the effects of smoke inhalation.
Four of the victims were located in the east wing, four in the
west, and one was located in the dining room.   Their ages were:

```
              78  (2)       89
              82  (2)       90
              84            97
              88
```

The Firefighters who removed the victims from the structure were
interviewed. They stated only 1 victim was found in bed. The body
positions of the others indicated they were attempting to escape
the building when they were overcome.

Also, 8 residents of the facility were treated for smoke inhalation
and transported to area hospitals. Their ages were:

```
              72            88  (2)
              77            91
              80
              83  (2)
```

Five Colorado Springs Firefighters suffered injuries and were
transported to Memorial Hospital for treatment.   The injuries
included:

      1 - Rib injury due to stepping into a window well and
          falling against its side (Paramedic)

      1 - Foot puncture from stepping on a nail during
          overhaul (Firefighter)

          The firefighter was wearing work boots at the
          time.

      1 - Smoke inhalation and exhaustion (Captain1

      1 - Pulled back muscles due to moving heavy file
          cabinets from the office area during property
          conservation efforts (Driver-Engineer)

      1 - Back injury due to falling while attempting to
          pull himself up into an attic space (Firefighter)

Rev. 3/21/91

Appendix B (continued)

Crystal Springs Estate
3/04/91   Page 7

Twenty-one fire personnel were evaluated by rehab members and were determined not to require further medical evaluation.

                    Captain   1
                    Lieutenants   6
                    Driver-Engineers   3
                    Firefighters   11

There were also 4 employees of private ambulance companies and 2 civilian rescuers/participants treated for smoke inhalation at local hospitals.

The structure suffered $500,000 to $750,000 in direct fire loss.

Photos of the fire's progress and of the investigation are on file at the Investigation Bureau.

Rev. 3/21/91

APPENDIX C

Alarm and Dispatch Log

```
FILE.NO LOCATION                              CALL TYP DESCRIPTION TIME RE

02072    825 S HANCOCK AV                     AUTOMATIC ALARM    00:35:40

UNIT  DISPATCH  RESPOND    ARRIVED    IN.SERV                    CALL.NO

E1    00:38:10  00:38:35   00:42:30                             033152
T1    00:38:12  00:39:12   00:43:21
E8    00:38:14  00:38:52   00:43:22
DC1   00:38:20  00:40:07   00:42:58
T8    00:43:49  00:45:21   00:49:51
E4    00:43:51  00:46:17   00:54:35   04:28:04
E2    00:43:54  00:46:19   00:50:57   08:10:35
RLT   00:47:16  00:47:24   00:52:14   07:42:44
DC2   00:49:17  00:49:23   01:00:17
E3    00:53:56  01:00:21   01:05:30   04:30:35
192W  01:01:53  01:02:01   01:02:03   06:00:47
E6    01:04:19  01:04:54   01:10:29   04:55:57
312   04:57:22  04:57:09   05:09:20   03:41:20
```

16

APPENDIX D

Uniform Fire Code, 1970 Edition

Crystal Estates was then classified as a "D."
Shows when the building would have had to be first sprinklered.

TABLE NO. 5-A-WALL AND OPENING PROTECTION OF OCCUPANCIES BASED ON LOCATION OF PROPERTY

TYPE IV AND V CONSTRUCTION: For Exterior wall and opening protection of Types IV and V buildings, see table below. Type V Construction is not permitted within Fire Zone No. I. Exceptions to limitations for Type IV and Type V Construction, as provided in Section 1109, 2103 and 5-A 2203 apply. For Types I, II, and III Construction see Sections 1803, 1903, and 2003.

GROUP	DESCRIPTION OF OCCUPANCY	FIRE ZONE	FIRE RESISTANCE OF EXTERIOR WALLS	OPENINGS IN EXTERIOR WALLS
A	Any assembly building with a stage and an occupant load of 1000 or more in the building		Not applicable [See Section 602 (a)]	
	1—Any assembly building with a stage and an occupant load of less than 1000 in the building	1	2 hour less than 20 feet / 1 hour elsewhere	Not permitted less than 5 feet / Protected less than 20 feet
	2—Any assembly building without a stage and having an occupant load of 300 or more in the building including such buildings used for educational purposes less than 12 hours per week or four hours in any one day and not classed as a Group C or Group F, Division 2 Occupancy	2 and 3	2 hour less than 10 feet / 1 hour elsewhere	Not permitted less than 5 feet / Protected less than 10 feet
B See also Section 702	3—Any assembly building without a stage and having an occupant load of less than 300 in the building, including such buildings used for educational purposes less than 12 hours per week or four hours in any one day and not classed as a Group C or Group F, Division 2 Occupancy	1	2 hour less than 20 feet / 1 hour elsewhere	Not permitted less than 5 feet / Protected less than 20 feet
		2	2 hour less than 5 feet / 1 hour elsewhere	Not permitted less than 5 feet / Protected less than 10 feet
		3	2 hour less than 5 feet / 1 hour less than 10 feet	
	4—Stadiums, reviewing stands, and amusement park structures not included within Group A nor Divisions 1, 2 and 3, Group B, Occupancies	1	2 hour less than 20 feet / 1 hour elsewhere	Protected less than 20 feet
		2	1 hour	
		3	1 hour less than 10 feet	Protected less than 10 feet
C See also Section 802	1—Any building used for educational purposes through the 12th grade by 50 or more persons for more than 12 hours per week or four hours in any one day	1	2 hour less than 20 feet / 1 hour elsewhere	Not permitted less than 5 feet / Protected less than 20 feet
	2—Any building used for educational purposes through the 12th grade by less than 50 persons for more than 12 hours per week or four hours in any one day	2	2 hour less than 10 feet / 1 hour elsewhere	Not permitted less than 5 feet / Protected less than 10 feet
	3—Any building used for day care purposes for more than 6 children	3	2 hour less than 5 feet / 1 hour less than 10 feet	

NOTES: (1) See Section 504 for type of walls affected and requirements covering percentage of openings permitted in exterior walls.
(2) For additional restrictions see Chapters under Occupancy, Fire Zones, and Types of Construction.
(3) For walls facing streets, yards and public ways see Part V.
(4) Openings shall be protected by a fire assembly having a three-fourths-hour fire-protection rating.

TABLE NO. 5-A—Continued

GROUP	DESCRIPTION OF OCCUPANCY	FIRE ZONE	FIRE RESISTANCE OF EXTERIOR WALLS	OPENINGS IN EXTERIOR WALLS
D See also Section 902	1—Mental hospitals, mental sanitariums, jails, prisons, reformatories, houses of correction, and buildings where personal liberties of inmates are similarly restrained		Permitted in Type I and II Buildings only [See Section 902 (b)]	
	2—Nurseries for full-time care of children under kindergarten age. Hospitals, sanitariums, nursing homes with nonambulatory patients, and similar buildings (each accommodating more than five persons)	1	2 hour less than 20 feet / 1 hour elsewhere	Not permitted less than 5 feet / Protected less than 20 feet
		2 and 3	2 hour less than 5 feet / 1 hour elsewhere	Not permitted less than 3 feet / Protected less than 10 feet
	3—Nursing homes for ambulatory patients, homes for children of kindergarten age or over (each accommodating more than five persons)	1	2 hour less than 20 feet / 1 hour elsewhere	Not permitted less than 5 feet / Protected less than 20 feet
		2 and 3	1 hour	Not permitted less than 3 feet / Protected less than 10 feet
E See also Section 1002	1—Storage and handling of hazardous and highly inflammable or explosive materials other than flammable liquids		Not permitted in Fire Zones Nos. 1 and 2	
		3	4 hour less than 5 feet / 2 hour less than 10 feet / 1 hour less than 20 feet	
	2—Storage and handling of Class I, II and III flammable liquids, as specified in U.B.C. Standard No. 9-1; dry cleaning plants using flammable liquids, paint stores with bulk handling; paint shops and spray painting rooms and shops	1	4 hour less than 20 feet / 1 hour elsewhere	
		2	4 hour less than 5 feet / 2 hour less than 10 feet / 1 hour elsewhere	Not permitted less than 5 feet / Protected less than 20 feet
	3—Woodworking establishments, planing mills and box factories; shops, factories where loose, combustible fibers or dust are manufactured, processed, or generated; warehouses where highly combustible material is stored	3	4 hour less than 5 feet / 2 hour less than 10 feet / 1 hour less than 20 feet	
	4—Repair garages		Not permitted in Fire Zones Nos. 1 and 2 except as set forth in Sections 1602 (c) and 1603 (c).	
	5—Aircraft repair hangars	3	1 hour less than 60 feet	Protected less than 60 feet
F See also Section 1102	1—Gasoline and service stations, storage garages where no repair work is done except exchange of parts and maintenance requiring no open flame, welding, or the use of highly flammable liquids	1	2 hour less than 20 feet / 1 hour elsewhere	Not permitted less than 5 feet / Protected less than 20 feet
	2—Wholesale and retail stores, office buildings, drinking and dining establishments having an occupant load of less than 100, printing plants, municipal police and fire stations, factories and workshops using material not highly flammable or combustible, storage and sales rooms for combustible goods, paint stores without bulk handling. Buildings or portions of buildings having rooms used for educational purposes, beyond the 12th grade with less than 50 occupants in any room	2	1 hour	Not permitted less than 5 feet / Protected less than 10 feet
		3	1 hour less than 10 feet	
	3—Aircraft hangars where no repair work is done except exchange of parts and maintenance requiring no open flame, welding, or the use of highly flammable liquids	1	2 hour less than 20 feet / 1 hour elsewhere	Not permitted less than 5 feet / Protected less than 20 feet
		2	1 hour	
	Open parking garages. (For requirements, see Section 1109.)	3	1 hour less than 20 feet	

(Continued)

Appendix D (continued)

CHAPTER 38 — FIRE-EXTINGUISHING SYSTEMS

Scope

Sec. 3801. (a) General. All fire-extinguishing systems required in this Code shall be installed in accordance with the requirements of this Chapter.

All hose threads used in connection with fire-extinguishing systems shall comply with the standards of the Fire Department.

(b) Approvals. All fire-extinguishing systems including automatic sprinklers, combination standpipes, dry and wet standpipes, special automatic extinguishing systems, and basement inlet pipes shall be approved and shall be subject to such periodic tests as may be required. The location of all Fire Department connections shall be approved by the Fire Department.

(c) Definitions. For the purpose of this Chapter, certain terms are defined as follows:

COMBINATION STANDPIPE is a fire line system with a constant water supply and installed for the use of the Fire Department and the occupants of the building.

DRY STANDPIPE is a fire line system without a constant water supply and equipped with Fire Department inlet and outlet connections and installed exclusively for the use of the Fire Department.

FIRE DEPARTMENT HOSE CONNECTION is a hose connection at grade or street level for use by the Fire Department only.

WET STANDPIPE is an auxiliary fire line system with a constant water supply installed primarily for emergency fire use by the occupants of the building.

(d) Standards. Fire-extinguishing systems shall comply with U.B.C. Standards No. 38-1 and No. 38-2.

Automatic Fire-Extinguishing Systems

Sec. 3802. (a) General. Automatic fire-extinguishing systems shall comply with the provisions of this Section.

(b) Where Required. Standard automatic fire-extinguishing systems shall be installed and maintained in operable condition as specified in this Chapter in the following locations:

1. In every story, basement or cellar of all buildings except Group I Occupancies when floor area exceeds 1500 square feet and there is not provided at least 20 square feet of opening entirely above the adjoining ground level in each 50 lineal feet or fraction thereof of exterior wall in the story, basement or cellar on at least one side of the building. Openings shall have a minimum dimension of not less than 30 inches. Such

---

openings shall be maintained readily accessible to the Fire Department and shall not be obstructed in a manner that fire fighting or rescue cannot be accomplished from the exterior.

When openings in a story are provided on only one side and the opposite wall of such story is more than 75 feet from such openings, the story shall be provided with an approved automatic fire-extinguishing system, or openings as specified above shall be provided on at least two sides of the exterior walls of the story.

If any portion of a basement or cellar is located more than 75 feet from openings required in this Section, the basement or cellar shall be provided with an approved automatic fire-extinguishing system.

2. Under the roof and gridiron, in the tie and fly galleries and in all places behind the proscenium wall of stages, over enclosed platforms in excess of 500 square feet in area; and in dressing rooms, workshops and storerooms accessory to such stages or enclosed platforms.

EXCEPTIONS: 1. Stages or enclosed platforms open to the auditorium room on three or more sides.

2. Altars, pulpits, or similar platforms and their accessory rooms.

3. Stage gridirons when side wall sprinklers with 135°F. rated heads with heat-baffle plates are installed around the entire perimeter of the stage at points not more than 30 inches below the gridiron, nor more than 6 inches below the baffle plate.

4. Understage or under enclosed platform areas less than 4 feet in clear height used exclusively for chair or table storage and lined on the inside with materials approved for one-hour fire-resistive construction.

3. In any enclosed usable space below or over a stairway in Groups B, C, and D Occupancies. See Section 3308 (f).

4. In basements or cellars larger than 1500 square feet in floor area of Groups A, B, and C Occupancies.

5. In all Group D Occupancies except jails, prisons and reformatories; however, the respective increases for area and height specified in Sections 506 (c) and 507 shall be permitted.

6. In Group E, Divisions 1 and 2 Occupancies having an area of more than 1500 square feet; in Group E, Division 3

1988 EDITION 9.116-9.117

2. Material having a structural base of non combustible material as defined in Item No. 1 above. with a surfacing material not over 1/8-inch thick which has a flame-spread rating of 50 or less.

"Noncombustible" does not apply to surface finish materials. Material required to be noncombustible for reduced clearances to flues. heating appliances or other sources of high temperature shall refer to material conforming to item 1. No material shall be classed as noncombustible which is subject to increase in combustibility or flame-spread rating beyond the limits herein estabiishcd. through the effects of age. moisture or other atmospheric condition.

Flame-spread rating as used herein refers to rating obtained according to tests conducted as specified in U.B.C. Standard No. 42-1.

NONFLAMMABLE MEDICAL GAS is a compressed gas which is nonflammable and used for therapeutic purposes and shall include, among others. oxygen and nitrous oxide.

NORMAL TEMPERATURE PRESSURE (NTP) is a temperature of 70°F and a pressure of 1 atmosphere 14.7 psi).

O

Sec. 9.117. **OSHA** is the Occupational Safety and Health Administration.

OCCUPANCY is the purpose for which a building or part thereof is used or intended to be used.

OCCUPANCY CLASSIFICATION. For the purpose of this code. certain occupancies arc defined as follows:

Group A Occupancies:

Division 1. Any assembly building or portion of a building with a stage and an occupant load of 1000 or more.

Division 2. Any building or portion of a building having an assembly room with an occupant load of less than 1000 and a stage.

Division 2.1 Any building or portion of a building having an assembly room with an occupant load of 300 or more without a stage. including such buildings used for educational purposes and not classed as a Group E or Group B. Division 2 Occupancy.

Division 3. Any building or portion of a building having an assembly room with an occupant load of less than 300 without a stage, including such building used for educational purposes and not classed as a Group B or Group B. Division 2 Occupancy.

Division 4. Stadiums. reviewing stands and amusement park structures not included within Group A Occupancies.

Group B Occupancies:

Division 1. Gasoline service stations garages where no repair work is done except exchange of parts and maintenance requiring no open flamc welding or the use Class I. II or III-A liquids.

35

Appendix E (continued)

Division 2. Drinking and dining establishments having an occupant load of less than 50, wholesale and retail stores, office buildings, printing plants, municipal police and fire stations, factories and workshops using materials not highly flammable or combustible, storage and sales rooms for combustible goods, paint stores without bulk handling.

Buildings or portions of buildings having rooms used for educational purposes beyond the 12th grade with less than 50 occupants in any room.

Division 3. Aircraft hangars where no repair work is done except exchange of parts and maintenance requiring no open flames. welding or the use of Class I or II flammable liquids.

Open parking garages.

Helistops.

Division 4. ice plants. power plants, pumping plants. cold storage. creameries.

Factories and workshops using noncombustible and nonexplosive materials.

Storage and sales rooms of noncombustible and nonexplosive materials.

Group E Occupancies:

Division 1. Any building used for educational purposes through the 12th grade by 50 or more persons for more than 12 hours per week or four-hours in any one day.

Division 2. Any building used for educational purposes through the 12th grade by less than 50 persons for more than 12 hours per week or four hours in any one day.

Division 3. Any building used for day-care purposes for more than six children.

Group H Occupancies:

Division 1. Occupancies with quantity of material in the building in excess of those listed in Table No. 9-A which present a high explosion hazard. including but not limited to:

1. Explosives. blasting agents. fireworks and black powder.

> **EXCEPTION:** Storage and the use of pyrotechnic special effect materials in motion picture television, theatrical and group entertainment production when under permit as required in the Fire Code, The time period for storage shall not exceed 90 days.

2. Unclassified detonatable organic peroxides.

3. Class 4 oxidizers.

4. Class 4 or Class 3 detonatable unstable (reactive) materials.

Division 2. Occupancies with quantity of material in the building in excess of those listed in Table No. 9-A which present a moderate explosion hazard or a hazard from accelerated burning. including but not limited to:

1. Class organic peroxides.

2. Class 3 nondetonatable unstable (reactive) materials.

Appendix E (continued)

3. Pyrophoric gases.

4. Flammable or oxidizing gases.

5. Class I. II or III-A flammable or combustible liquids which are used in normally open containers or systems or in closed containers pressurized at more than 15-pounds-per-square-inch gage.

6. Combustible dusts in suspension or capable of being put into suspension in the atmosphere of the room or area.

> **EXCEPTIONS:** 1. Rooms or areas used for woodworking that do not exceed 500 square feet in area may be classified as Group B. Division 2 Occupancies. provided dust-producing machines are equipped with approved dust collectors and there are not more than two such machines.
>
> 2. Lumberyards and similar retail stores utilizing only power saws may be classified as Group B. Division 2 Occupancies.

The building official may revoke the use of these exceptions for due cause.

7. Class 3 oxidizers.

Division 3. Occupancies with quantity of material in the building in excess of those listed in Table No. 9-A which present a high fire or physical hazard. including but not limited to:

1. Class II, III or IV organic peroxides.

2. Class 1 or 2 oxidizers.

3. Class 1. II or III-A flammable liquids or combustible liquids which are utilized or stored in normally closed containers or systems and containers pressurized at 15-pounds-per-square-inch gage or less.

4. Class III-B combustible liquids

5. Pyrophoric liquids or solids.

6. Water reactives.

7. Flammable solids. including combustible fibers or dusts, except for dusts included in Division 2.

8. Flammable or oxidizing cryogenic fluids (other than inert).

9. Class 1 or 2 unstable (reactive) materials.

Division 4. Repair garages not classified as a Group B. Division 1.

Division 5. Aircraft repair hangars and heliports not classified as Group B, Division 3.

Division 6. Semiconductor fabrication facilities and comparable research and development areas when the facilities in which hazardous production materials (HPM) arc used and the aggregate quantity of materials are in excess of those listed in Table So. 9-A or 9-B. Such facilities and areas shall be designed and constructed in accordance with Section 911.

Division 7. Occupancies having quantities of materials in excess of those listed in Table No. 9-B that are health hazards. including but not limited to:

1. Corrosives.

2. Highly toxic materials

Appendix E (continued)

9.117   UNIFORM FIRE CODE

3. Irritants.

4. Sensitizers.

5. Other health hazards.

Group I Occupancies:

Division 1. Nurseries for the full-time care of children under the age of six teach accommodating more than five persons).

Hospitals. sanitariums. nursing homes with nonambulatory patients and similar buildings (each accommodating more than five persons).

Division 2. Nursing homes for ambulatory patients. homes for children six years of age or over teach accommodating more than five persons).

Division 3. Mental hospitals. mental sanitariums. jails. reformatories and buildings where personal liberties of inmates are similarly restrained.

　　　　EXCEPTION: Group I Occupancies shall not include buildings used only for private residential purposes or for a family group.

Group M Occupancies:

Division 1. Private garages. sheds and agricultural buildings when not over 1000 square feet in area.

Division 2. Fences, tanks and towers.

Group R Occupancies:

Division 1. Hotels and apartments. Convents and monasteries (each accomo. dating more than 10 persons).

Division 2. Not used.

Division 3. Dwellings and lodging houses.

OIL-BURNING EQUIPMENT is an oil burner of any type together with its tank. piping, wiring, controls and related devices and shall include all oil burners. oil-fired units and heating and cooking appliances but exclude those exempted by Section 61.101.

OIL-FIRED UNIT is a heating appliance equipped with one or more oil burners and all the necessary safety controls, electrical equipment and related equipment manufactured for assembly assembly complete unit. This definition does not include kerosene stoves or oil stoves.

OPEN-AIR GRANDSTANDS and BLEACHERS are seating facilities which are located so that the side toward which the audience faces is unroofed and without an enclosing wall.

OPERATING LINE is a group of separated operating buildings of specities arrangement used in the assembly. modification. reconditioning, renovation. maintenance. inspection, surveillance, testing or manufacturing of explosives.

ORGANIC COATING is a liquid mixture of binders such as alkyd, nitrocellulose. acrylic or oil and flammable and combustible solvents such as hydrocarbon. ester. ketone or alcohol. which when spread in a thin film converts to a durable protective and decorative finish.

38

APPENDIX F

Fire Prevention Inspection Log Records for 1990 and 1991

```
.    2 LINE(S) FOUND OUT OF 1677 LINES
.
* * * * * * * * * * * * * * * * * * *  *  *****
*
* HANCOCK AV              S  00825
.
:DATE          10:11:06   RID      9G   07 JAN 91  HUAN
.@921231                      FIRE PREVENTION INSPECTION LOG          G0004.
*RECS FROM 10/17 ARE HRE PREVIOUS RECS IN 10/88 ARE IN 46G      6G
*STREET NAME          :D:ST-NUM    :U-TYP:U-NUM:SC:ACT:HOURS:A-DATE:INIT:R-DA:
=============.=.========.=======.===.===.=====.======.====.======
HANCOCK AV            S 000825              06 IA   2.50 880630 PJR
.
.       ******  C O M M E N T S  *** BY : PJR ON : 880630 RPT 88019
.
.001= 10.308 REMOVE PAINT FROM HEAT DETECTORS W/REQ.
.002= 10.402 ADJUST ALL EXTERIOR EXIT TO OPERATE PROPERLY-
.003= 10.402 ADJUST INTERMEDIATE CORRIDOR TO OPERATE PROPERLY AND TO LATCH.
.003= 10.402 COMPLETE THE INSTALLATION OF THE MAGNETIC HOLD OPEN DEVICE AT
.005= THE KITCHEN DOOR.
.006= 10.314 INST A 40-B FR. EXT. IN KITCHEN ADJACENT TO THE COOKING AREA.
.007= 12.101 INST GUARDRAILING ADJACENT TO THE NORTH EXIT DOOR FROM DINING RM.
.008= 10.302 MOUNT EXISTING FR. EST. ON BRACKETS.
.009= 10.301A REPLACE 1-A FR. EXT. IN THE BASEMENT.
.010= 10.401 ACHIEVE COMPLETE 1 HR. FR. RESISTIVE CONST. IS THE BASEMEST MECH.
.011= 85.106 REPLACE TEMPORARY ELECTRICAL WIRING IN THE WEST CRAWLSPACE.
.012= 85.104 INST. METAL COVER PLATES OVER ELECT. JUNCTION BOX IN THE EAST CRAW
.013= 12.103 MAKE SOUTH EXIT DOOR FROM BASEMENT COMPLETELY OPERATIONAL.
.01d= SER. KITCHEN FR. EXT SYS. TAG SAME.
.015= 12.114 MAKE EXIT SIGNS IN THE WEST AND NORTH WING FULLY FUNCTIONAL.
.016= 11.2038 REMOVE ALL COMBUSTABLE WASTE AND DEBRIS.
.017= 25.107 MAINTAIN 48" CLEAR AISEL MAIN CORRIDOR.
.018= 25.107 MAINTAIN 36" CLEAR ASILE IN THE ROOMS OFF THE MAIN CORRIDOR.
.019= 12.103E MAINTAIN 24" CLEARANCE FROM CEILINGS TO COMBUSTABLE.
==================================================================
HANCOCK AV            S 000825              06 IA   0.50 880719 PJR
.
.     ****  C O M M E N T S  *** BY : PJR OS : 880719 RPT 880215
====================================================================
.
:001 = REINSP. APPD.      . . . . . . END REPORT . . . . . .
==============================================================
```

23

Appendix F (continued)

DATA ENTRY

ATE O 82990

CITY OF COLORADO SPRINGS
FIRE DEPARTMENT
INSPECTION ASSIGNMENT

PAGE 1 OF

A. FDZ	STREET	D	NUMBER	UNIT TYPE	UNIT NO.	DATE ISSUED	V	PRI.	INSPECTIO
01 0489	HANCOCK AV	S	825			02/05/90	N	3	25254

1-USE	2-OCC. DESCRIPTION	3-OCC NAME	4-OCC. PHONE
310	CARE OF THE AGED	CRYSTAL SPRING ESTATE	630 1441

5-BUSINESS OWNER/MANAGER	6-PHONE	7-PROPERTY OWNER	8-PHONE (INCL. AREA CODE)
SANNER JIM	599 8852	SANNER JIM	719 599 8852

9-INSURANCE COMPANY	10-AGENT	11-PHONE (INCL. AREA CODE)
USF&G BALTIMORE MD	PEGGY	303 534 2133

12-TOTAL SQ. FEET	13-RESPONSIBLE PARTY	14-PHONE
12,600	SANNER JIM	599 8852

14-STRUCTURAL REMARKS
PEAK SECURITY MONITORS AND REPAIRS ALARM SYSTEM 630-1200

15-STRUCTURAL REMARKS

16-LOCK BOX LOCATION	17-WATER SHUT-OFF LOCATION	18-TEST DATE	19-FIRE ALARM PANEL LOCATION
N/A	S BASEMENT	08/87	NURSE'S DESK

20-GAS SHUT-OFF LOCATION	21-ELECTRICITY SHUT-OFF LOCATION	22-TEST DATE	23-STAND PIPE CONNECTION LOCATION
SW CORNER	BEHIND NURSE STA	N/A	N/A

24-SPRINKLER CONNECTION LOCATION	25-SPRINKLER AREA SERVED	26-TEST DATE	26-STAND PIPE CONNECTION LOCATION
NEXT TO FRONT DOOR	BASEMENT	08/87	N/A

24-SPRINKLER CONNECTION LOCATION	25-SPRINKLER AREA SERVED	27-EXTINGUISHER SYSTEM AREA SERVED
FRONT DOOR WEST SIDE	KITCHEN & PANTRY	KITCHEN HOOD

24-SPRINKLER CONNECTION LOCATION	25-SPRINKLER AREA SERVED	27-EXTINGUISHER SYSTEM AREA SERVED
N/A	N/A	N/A

28-EMERGENCY REMARKS
OXYGEN STORAGE-DINING RM E WALL.W WING STORAGE EXCEPT FOR 1 RESIDENCE S END

	29-NO.	30-HAZARD LOCATION	31-HAZARD AMOUNT	32-HAZARD DESCRIPTION
L				
L				
L				
L				
L				
L				

Appendix F (continued)

CITY of COLORADO SPRINGS
FIRE DEPARTMENT
INSPECTION REPORT

PAGE ___1___ OF ___1___

STA	STREET		D	NUMBER	UNIT TYPE	UNIT NUMBER	INSPECTION

OCCUPANT NAME		OCCUPANT PHONE
Crystal Spring Estate		*6-0 1441*

BOXES CHECKED INDICATE VIOLATON   ☒ **NO VIOLATIONS NOTED**

	CORRECT BY DATE	OK

A. General Hazards

☐ Address Numbers Not Visible, Legible From Street (5" mininum standard)   10.208 _____ __

☐ Hazardous Accumulation of Waste, Rubbish, Etc.   11.201 _____ __

B. Exits, Passage Ways, Life Safety

☐ Egress Doors Locked, Blocked, Inoperative, "This Door to Remain Unlocked During Business Hours" Sign Needed   12.104 _____

☐ Corridors, Passage Ways, Stairways Obstructed   12.103 _____

☐ Storage Under Exit Stairways   12.106 _____

☐ Exit Signs Not in Place, Not Maintained   12.114 _____

☐ Required Self-Closing Doors Inoperable, Obstructed   12.104 _____

C. Flammable Liquids

☐ Not Properly Stored or Handled   79.201 _____ ✓

D. Fire Protection Systems, Appliances

☐ Fire Extinguishers Not Maintained, Insufficient Number, Not Mounted   10.302 _____

☐ Cooking Hood Systems Require Cleaning, Serviced, Inspected   10.314 _____

☐ Smoke/Heat Detection, Alarm Systems Tested Annually   10.302 _____

☐ Sprinkler System/Standpipe Not Maintained, Not Inspected, Inaccessible   10.302 _____

E. Flame Producing Appliances

☐ Combustibles Stored Too Close to Appliance   11.203 _____

F. Electrical Hazards

☐ Improper Wiring, Extension Cords thru Walls, Ceilings, Across Travel Paths   85.104 _____

G. Others

_____

_____

Rechecked By_____

The items noted above are in violation of the Uniform Fire Code, as amended. This is an official notice of code violations requiring correction within the specified time. Failure to comply with these requirements may lead to legal action. Violation of the Fire Code is a misdemeanor punishable by a fine not to exceed $500.00 or imprisonment not exceeding 90 days, both. This inspection is intended for your safety and the safety of the citizens of Colorado Springs. Your cooperation is greatly appreciated. For information concerning this inspection call 578-7040.

COPY RECEIVED BY	DATE	INSPECTED BY	INT.	TITLE	DUTY ASSIGN
	7-29-90	*McClendon*			

NO INSPECTORS	TIME IN	TIME OUT	DATE OF INSPECTION	REFER TO FIRE PREVENTION		CODE REVISION
	1430	1630		☐ YES		☒

Photographs

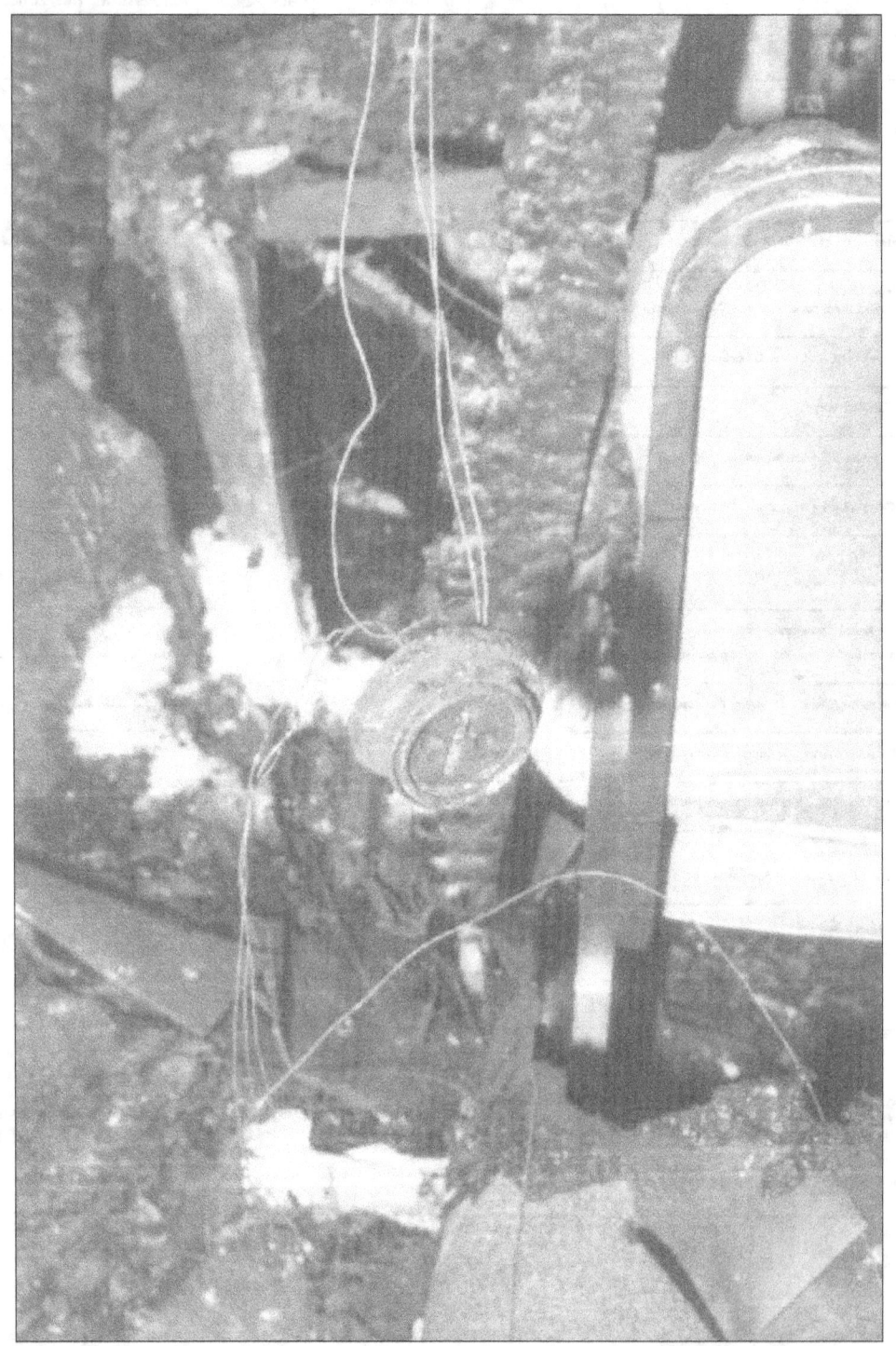

Photo by Jack Yates

Heat sensors from the attic, either not hooked up or not working as there was no early warning to any panel.

Appendix G (continued)

Exterior of the structure as seen from the southeast and looking to the northwest. The east wing is to the right and the south wing is to the left.

Appendix G (continued)

Photo by Jack Yates

Exterior as viewed from the south and looking north at the wall and roof area over the dining room.

Appendix G (continued)

Photo by Jack Yates

View of the east wing hall as seen from the south to the north. The smoking room where the fire began overhead is the first door on the left.

Appendix G (continued)

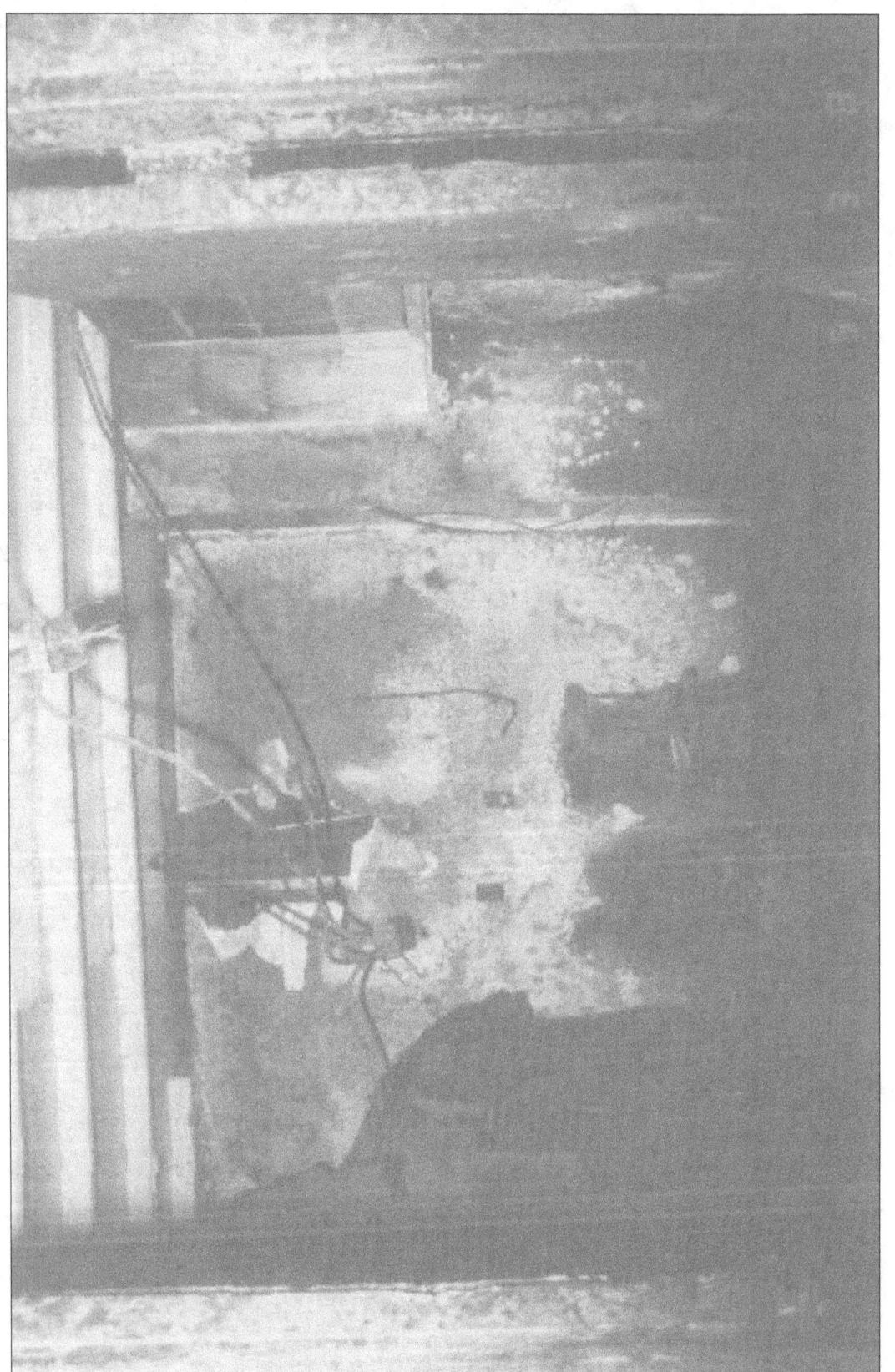

Photo by Jack Yates

The smoking room. The corner to the right is the northwest corner where the vent stack to the heating unit below is located. This was thought originally to be the Area of Origin.

Appendix G (continued)

Photo by Jack Yates

View of the northeast corner of the smoking room. The light switch behind the door (just under the white spot in the center) controlled the fluorescent lights and the ventilation fan. The fan was located in the northeast corner. When the lights were on the fan was on. Since the switch location was difficult to get to, the lights and fan stayed on for hours and sometimes days at a time.

Appendix G (continued)

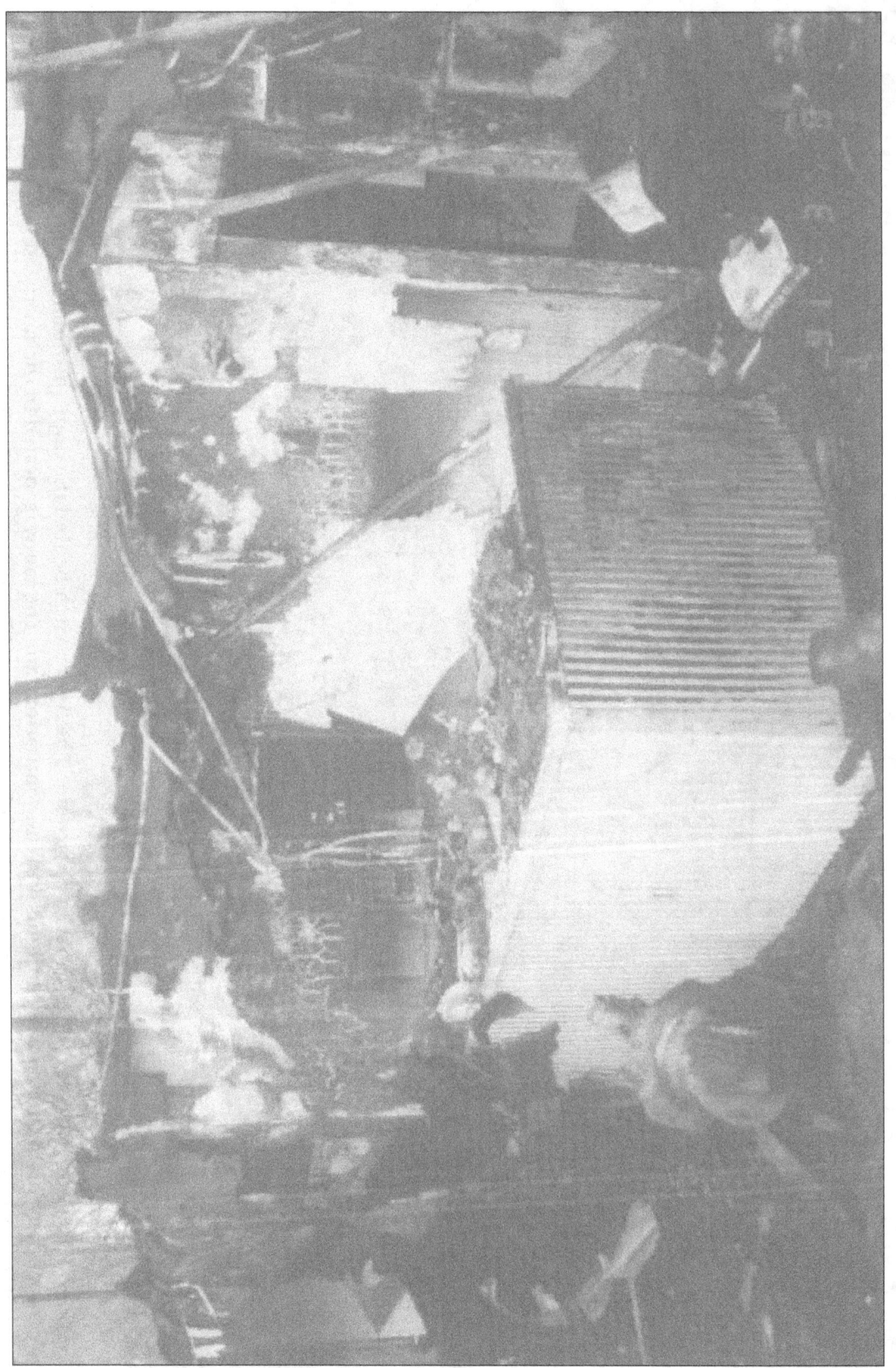

Nurses' station just inside the front (southwest) door. The alarm panel is at this station. View is toward the northeast.

Photo by Jack Yates

Appendix G (continued)

Photo by Jack Yates

Looking north from the entry area by the nurses' station down the hall. Note all overhead framing was destroyed.

Appendix G (continued)

Photo by Jack Yates

Front door through which initial interior attack was made, located at the southwest corner of the structure. View is from the interior toward the west.

Appendix G (continued)

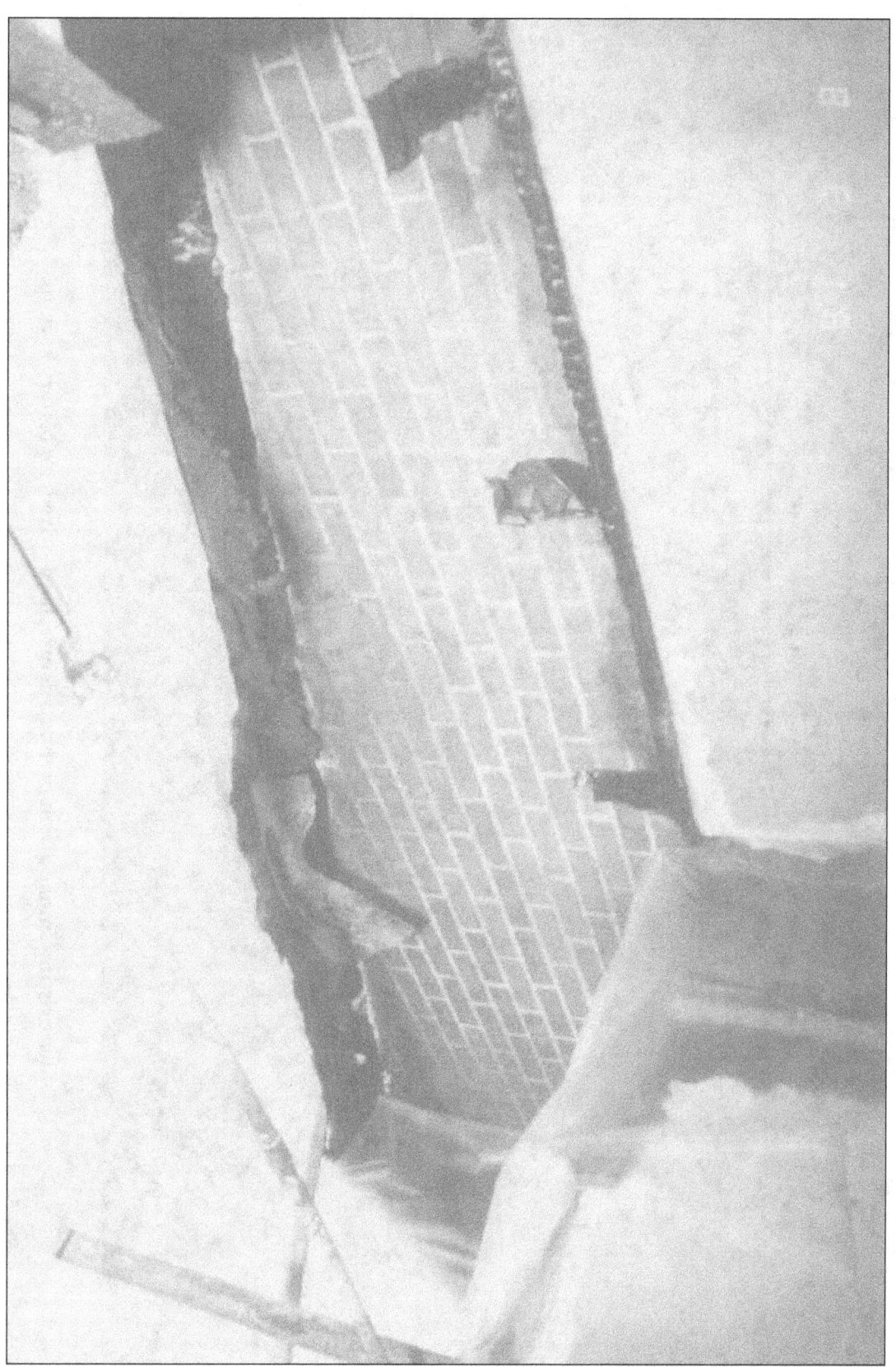

Photo by Jack Yates

The east wall of the dining area. Note where holes have been punched in to put joists through. There was a cinder block wall behind this, but tongue-in-groove boards extended over the top exposing the boards to the brick area. Integrity of the firewall was breached from the holes and overlap of wood material.

Appendix G (continued)

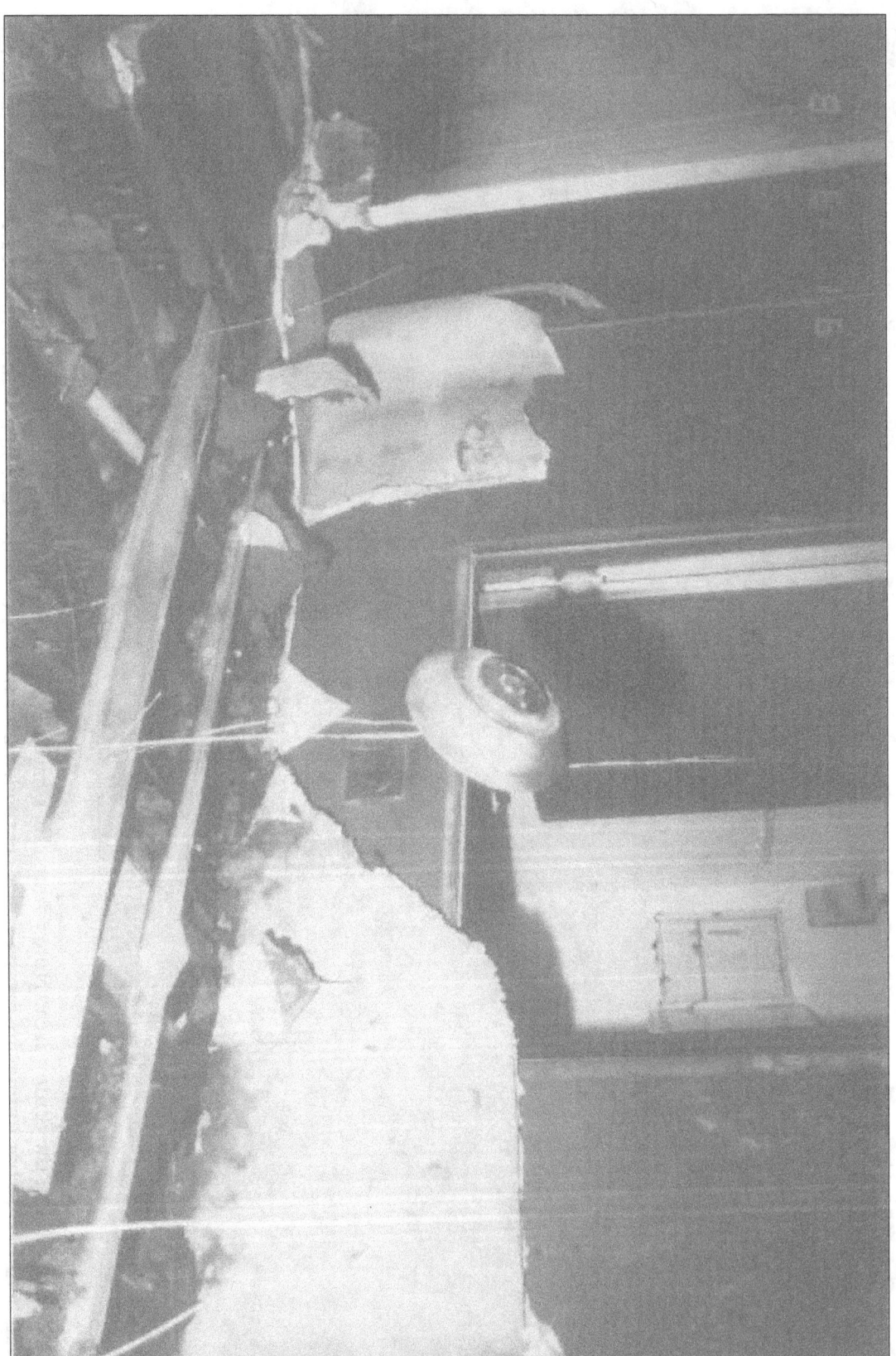

Photo by Jack Yates

Older existing smoke alarm in the structure, hard-wired, no battery backup.

Appendix G (continued)

Photo by Jack Yates

New smoke alarm installed approximately one week prior to the fire in order to be in compliance with State regulations.

www.ingramcontent.com/pod-product-compliance
Lightning Source LLC
Chambersburg PA
CBHW081237170526
45165CB00009B/3089